Copyright © 2019 by El Moutadir Hamza

This edition is published by Lulu Digital, by arrangement with EL Moutadir Hamza

All rights reserved. No part of this book may be used or reproduced in any manner whatsoever without the written permission of the Publisher.

2019 Lulu Digital

ISBN 978-1-79478-621-9

Terms and Conditions
LEGAL NOTICE

The Publisher has strived to be as accurate and complete as possible in the creation of this report, notwithstanding the fact that he does not warrant or represent at any time that the contents within are accurate due to the rapidly changing nature of the Internet.

While all attempts have been made to verify information provided in this publication, the Publisher assumes no responsibility for errors, omissions, or contrary interpretation of the subject matter herein. Any perceived slights of specific persons, peoples, or organizations are unintentional.

In practical advice books, like anything else in life, there are no guarantees of income made. Readers are cautioned to reply on their own judgment about their individual circumstances to act accordingly.

This book is not intended for use as a source of legal, business, accounting or financial advice. All readers are advised to seek services of competent professionals in legal, business, accounting and finance fields.

You are encouraged to print this book for easy reading.

Table of Contents

Chapter 1:

Introduction To Personality Typing.

Chapter 2:

How To Determine Your Personality Type

Chapter 3:

Personality Typing Tools

Chapter 4:

The Characteristic Of A Good Personality Type

Chapter 5:

The Benefits Of Personality Typing

Chapter 6:

The Factors To Be Considered In Personality Typing

Chapter 7:

The ABC Of Personality Typing

Chapter 8:

Learn To Understand Other Personality Types

Chapter 1
Introduction
Synopsis

Personality typing or types refer to the classification of the different types of individual psychologically. When you say personality types, it is the different kinds of personality or traits that a person or individual has. But when you say personality typing, it is the act of knowing the kind of personality that this person or individual portrays. It is easy to know the types of personality of a person through the use of the personality typing tools. These tools are used to determine the right personality of a person. Sometimes, the personality type can be distinguished through the personality traits of a person or by grouping the behavioral tendencies.

According to some experts, it is important that you know the personality of the person so that you will be able to make certain adjustments to the things that will make them feel comfortable or you can find a better spot for them in your company. For example, if you are hiring an employee, the personality test will let you know if this person is friendly, strict, hard working or deserving for the position that you are looking for. Personality typing can be given to anyone who wants to know the type of personality a single person has. Knowing the personal types of the people is not used to degrade them but it is used to know the capabilities of a person regarding their work.

If you don't know the personality of the people around you or within your workplace, you will surely find it hard to fit in or understand their world. This is because every person around you has his or her own different personality that you should understand. You will not find it hard to know their personality type once you get to know them well. Some people base their personality type on the experiences that they have encountered with the person. As you get to know the person, you will also know the type of personality that he or she has. Just make sure that you know the different types of personalities so that you will be aware of how to treat other people depending on their own personality.

You will find a lot of benefits in knowing your personality type and these benefits will surely help you move forward to your future. It will also help you gain more knowledge and ideas on how you can interact with other people. Personality typing is created to help everyone in their needs. If you don't know how to type your personality well, you will also not learn how to have a good relationship with other people around you.

Typing your personality helps you in improving and developing your self confidence and trust in yourself. Another important thing is that it will also help you motivate yourself more. Being in a world that is filled with many different personalities can be very difficult because things could get more complicated. People will argue a lot and a lot of misunderstandings and conflicts will happen. This is what people are trying to avoid, so you should know the tools that will help you understand the different types of personality that every person has.

Chapter 2:
How To Determine Your Personality Type
Synopsis

There are several ways on how you can type your personality. The very first thing that you can do is through some personality test. The personality test will help you determine the type of personality that an individual has. The term type is not used again and again in psychology and it has become the main source of confusion. Moreover, since the personality test score usually falls on the bell curve than the distinct categories, the personality type theories have received considerable criticism among the psychometric researchers.

The psychologist is the right person that you can ask on how you can type a person's personality. But since you are not a psychologist, there are also alternative personality tools that you can use to type a person's personality.

One of the things that you can use to type a personality is through the personality tests and questions. The personality tests will help you determine the type of a person easily. All you need to do is provide a simple test and some questionnaires that will test a person's ability, patience, flexibility and mind. As you go through the process of the personality tests, you will know the type of personality you have.

If you are flexible enough, patient and you can take all the personality tests that are provided to you, then at the end of the day, you can type your own personality. It is easier to type your personality because you already know yourself more than anyone else.

You can also try to test your personality type using different tools like doing exercise, visualizing and answering assessment tests that try and test your personality. Through these tools, you will be able to test your patience, your skills and your ability to handle the situation. Some people find it hard to type their personality and this is because others portray a dual personality. Sometimes, they are type A which is the reason why they are very aggressive. At times, they are type C which is the reason why they are so emotional.

If you are not yet sure about your personality, you can do some tests to know your real type of personality. Personality typing is done to yourself when you are not sure of your personality.

The Personality Test Center has a lot of questionnaires that can be used by anyone who wants to know his or her personality type. You will realize that some people don't work, think, talk or act like you. It is easy to say that your friends or co-workers are crazy, lazy or disorganized when you notice that their behavior is not matching your expectations. This will help you realize that you need to understand the personality differences of each person. It is just normal to see people act at their own age with different attitudes. There have been a lot of theories that say the human behavior is merely not random but classified and predictable. According to one of the theories, all people are born predisposed in certain personality preferences.

The personality typing helps you know and understand clearly the type of personality you have. You can easily identify your personality if you are guided accordingly from your experiences and how you are raised. The people that influenced you will also serve as a factor in identifying your personality type. These are the basis of your personality type.

Chapter 3:
Personal Typing Tools
Synopsis

The personality typing tools can help you determine the personality type of a person. These tools are not machines but they can be a very powerful in order to know the personality of a person. Knowing one's personality is so important especially if you are planning to hire a person or you are planning to build a strong relationship with him or her. Some of the tools that you can use for your personality typing test are the pen, paper and questionnaire.

This test will help you determine the type of your personality. For some people, their basic tool for determining their personality is based on their experience itself. Experience also helps people know what kind of person they are because personality is often learned from experiences.

Other personality typing tools that can help you determine your type are determination and visualization. Just writing down your goals will never give you any progress because you need to put it in action so that you will reach your goals. For example, you want to stay sexy and fit but you are not doing anything to help your body achieve that. What you need is determination with proper exercise.

If you are the type of person that is lazy and has no determination, then you will surely never achieve your goal. If you can visualize the things that you need to do, then you will find it easy to do the necessary actions. If you want to type your personality, then you should start doing what you have started or else, you will end up on that personality that is negative and not wanted by people.

Another tool that you can use to help you understand the personality types is motivation. You should be more motivated in learning how to recognize your personality and that of other people. Motivation is a good tool in learning and understanding the personality type of people.

Chapter 4:
The Characteristic of Good Personality Type
Synopsis

According to some theories, people have one of the six characteristics of a personality type such as realistic, enterprising, social, artistic, conventional and investigative. Below, you will find the description of each of the characteristics of a good personality type.

Realistic – They love to work with tools, animals or machines and they avoid social activities like teaching and giving information to others. They have good skills with the animals, tools, plants, machines and other electrical and mechanical drawings. The realistic people see things in a practical and realistic way and they also value certain things that you can feel, see and use. These are the people who know that if they motivate themselves, they can go wherever they want to be.

Investigative – This type of person is good in solving and understanding math and science problems. They love to study and solve problems alone and they also avoid selling, leading and persuading people. An investigative personality type values science and sees himself or herself as scientific, intellectual and precise.

Social – If you have the characteristic of a social personality, then you like to do things in order to help other people. You like teaching, giving first aid or nursing, providing information to all people and you don't want to use machines, animals and tools just to achieve your goal. You are also good in counseling and solving other peoples' problems because you see yourself as friendly, trustworthy and helpful to others. Being social is helping others in their needs whether they are physical, mental, spiritual, financial or moral needs.

Enterprising – An enterprising personality loves to persuade, lead other people and sell ideas and things. They also avoid people and activities that love to do careful observation, analytical thinking and scientific studies. They give important value to success in leadership, politics, business and other things because they see life as ambitious, sociable and energetic.

Conventional – The people with a conventional characteristic like to work on numbers, machines in set, and records and they also avoid people that are ambitious and have unstructured activities. They are also known to work well with written numbers and records in a systematic way. This kind of person values business so much and he or she sees himself or herself as orderly and follows the set plan accordingly.

Artistic – If you are an artistic person, you like to do any kind of creative activities such as crafts, drama, art, music, dance and creative writing. They avoid people with highly repetitive and ordered activities. They have the ability to create different kinds of creative activities. They surely give high value in this type of activities.

These are the important characteristics of a good personality. If you know your personality, then you can deal with other people properly. These characteristics are not by choice but they are adopted through the people around you and how you are raised. You can only choose to use your personality in a good or bad manner. If you will use it in a good manner, then you are sure to have a good future and be successful in your career.

Chapter 5:
The Benefits of Personality Typing
Synopsis

Everyone can surely benefit from knowing their personality type. Once you know the type of personality of a person and you try to understand them, you will become more effective, genuine and efficient. When the type training is implemented in a work place, the team begins to work as a group but as they try to cope with each other, you will notice that some people are not getting along with other groups well. This is because people are born with different personality types. So, it is also important that you know a way of identifying the different types of personality so that you can get along well with people around you. Your company will become more profitable, productive and attractive to the customers if you know how to deal with your workers.

Personal typing means knowing your personal type because it is important that you know it and the types of people around you. This gives you a heightened self-awareness and a lot of benefits. For sure, you will be happy with the benefits that you will get because it can also help you improve and become a better person. Sometimes, you will wonder why you are not like other people and why you cannot be like them. The answer is because you have a completely different personality type.

If you know exactly your type of personality, then you can be sure to have a good relationship with other people. You will know exactly how to treat people and give them the advice that they need. You can socialize with them and avoid any kind of misunderstandings.

- Feel good and accept yourself for who you really are. Most people are happy about their type and are so proud of it; these people see the nice aspects of their selves. They are very aware of their responsibilities and duties because they know the type of personality that they have. They feel good because they are not afraid to show other people who they really are and they can get along well with other individuals.

- They can identify their natural strengths. Everyone can do their best when working with other innate strengths. When you work with your strengths, it also gives you the enjoyment and energy. People can identify their weakness and blind spot. Just like a coin, each personality has its own flip side. If you have the strength, for sure, you also have your weakness. When you work with your weaknesses, it can drain your energy and sometimes cause frustrations. When you are caught in a situation that hits your blind spot, you tend to act and speak in a way you regret. This is why you should know your blind spot so that you can control the situation and you can do the right things.

- Choose a career that satisfies you whether it is an actual work or a work for an environment that means so much to you. Whatever makes you feel happy, safe and contented will surely give you the self satisfaction that you need. If you know your personality type, then you can easily choose the right career that you know you can be good at. You can find the right career with a healthy environment that you can surely fit in and use your own knowledge.

- You can develop better relationships by being understanding and accepting that people are different from you. For sure, you can be grateful for their differences when you realize that their perceptions and strengths can be of help to you. But this doesn't mean that you should stay away from those people with a negative personality. Instead, you should understand them more and help them become a better person.

These are the benefits you can get when you know how to type your personality. It is only you that can realize it and you just need to open up your mind for possibilities. It is never too late to know your personality because it will help you become happy and have a clear future ahead.

You should know yourself more than anyone else in the world so that no one can ever put you down and you will be able to take all the challenges in life. These benefits will be your guide, so make sure that you already know your personality type so that you can deal with other people properly.

Choosing a career that would best suit your personality is so important because if you are doing what you love, then surely, you will be happy and more satisfied with your achievements.

It would be so comfortable and happy to work if this is really what you love to do.

Chapter 6:
The Factors to Be Consider In Personality Typing
Synopsis

There are so many things that you need to consider in determining a personality type. It is important that you consider knowing your personality type so that you know how to deal with other people. In this article, you will find all the factors that you need to keep in mind in determining a personality type. The very first factor is acceptance. If you can accept the personality of the person, you will know how to deal with him or her properly. If you cannot accept what you really are, then you will also find it hard to accept your own personality. You will also find it hard to accept the personality of other people.

Personality type really matters because it gives an impact on everything that you say, do and think. It impacts the group dynamics that help form the culture of the organization. When you know how to handle your personality, you will also be responsible in all your acts, especially when you are under stress or during times of challenges. It also impacts how the people mentor, negotiate, manage, lead and solve problems. This can influence the time management behavior, goal setting, motivation and attentiveness in the meetings. It can even impact how you practice and approach spiritually.

You can also improve faster and easily if you can understand yourself more by simply addressing the right behaviors and thoughts. You can also act from outside your personality preferences if this is what the situation needs. As long as you know what you are doing, then you are free to do everything that makes you happy. There is a room for a big improvement on both sides of blind spots and natural preferences. Knowing the personality type will give you the insight that you need in order to develop your personal growth plan.

If you want to have a good relationship with your friends and loved ones, it is important that you know the things that you need to consider. This will also help you develop good characteristics and ensure your success in the career path you have chosen to take. You will also become a good problem solver, trainer, coach and leader. All in all, you will enjoy the life that you have always wanted. All the things that you need to achieve will be easier for you to reach because you know your personality type. This is why everyone is encouraged to know the different types of personality including your personality type.

With all the things that you need to consider, you must never forget to know the tools that will help you know your personality type. These tools will be your guide in order to have a good relationship with other people. You must always know the right types of people so that you will easily know their weaknesses and strengths.

Chapter 7:
The ABC Of Personality Typing
Synopsis

The personality typing can be characterized as ABC wherein each of the letters has a different meaning. If you have the type A personality, then you are most likely to suffer from heart diseases. This theory is based on the characteristics that all type A persons have. The traits include unnecessary obsession with proper time management, stress, hostility and aggression. Because of the stress factor, the researchers have found that the persons who have a type A personality are at risk of suffering from heart diseases. However, this theory is still being proven, but this is how their tests show.

For the people with the type B personality, they tend to have less stress and a slower pace than the stress of the type A personalities. This type of people is more tolerant and flexible and they can also overcome feelings of guilt so that they can take their time to enjoy their life. As the years passed by, the theory of the personality became widely accepted and is now used as a tool in understanding the personality of a person. This is why most of the professional careers these days are for the people with the type B personality. The tests and tools for personality checking can determine whether you are a type B or not. The psychologists are experts when it comes to knowing and understanding the personality types of a person. Through self-awareness and pre employment screening, employers will know the specific type of personality a person has.

The modern day psychologist added the type C personality as they can also see people with this kind of personality that is far from the type A and B. The type C has the same traits as the sub types but people who have this personality are more emotional. These types of persons are so emotional and they can get easily carried away by their emotions resulting to low self-esteem and other low traits. The personality type C is often mistaken with type A but the big difference is that they are not so obsessed with time management. The type C personalities are afraid to show and share their emotions to other people. They are also categorized as the emotionally repressed people.

You need to make sure that you know the different type of personalities so that you can deal with them properly. If you are hiring an employee, it is important that you know what you are looking for so that you will not have regrets in the end. You can try the personality assessment tests to know the personality type of the people that you are hiring. This way, it will also be easier for you give your trust and time to them. A lot of people are not hired because of personality problems. During the interview, you are also typing your personality.

If you are arrogant, bossy and loud or anything that is not allowed in a company, then you will most likely not get accepted in the company where you are applying.

Chapter 8:
Learning To Understand Personality Types
Synopsis

It is very important that you understand the different personality types of a person. If you want to get well with others, you must know that people are not all born the same, but equal. They may be different from each other but they are equal in terms of the different personalities they have. If you will not learn to understand the other personality types, then you might have difficulties in handling the different kinds of people. There are so many ways on how you can learn to understand the different types of personalities. The very first thing to do is to identify the type of personality of a person.

If you will not understand the different personalities of people, you will surely find it hard to deal with them. You will always be struggling to understand what they are trying to express to you. This is also the reason why there are so many people who are always having an issue because of misunderstanding and personality problems. One should always understand the personality of other people so that they will not have conflicts or misunderstandings. If you have the knowledge about the preference and personality types, then it will surely be useful to you. Through the personal typing, you will have the chance to do the right things that will make your future successful and happy.

If you know your personal type, then you can get along well with your co-workers; you will know your strengths and your weaknesses and you will understand everyone around you. But if you don't understand your personal type, then you will surely find it hard to deal with your co-workers. Every day, you would feel like you are in a fighting area where you always compete with your own ghost just to understand yourself and everyone around you.

Learning the personal type will give you so many advantages that can surely help you in your needs. A person's personality can also be used as a resource and tool to navigate your creative direction. You need to understand other people because if you cannot manage these types of personalities, it could become a big deficit in your project.

If you learn how to understand the different personality types of people, then you can use this as a motivation for yourself. This could help you motivate yourself and do good things to other people and yourself as well. You can also give a purpose to yourself and other people. Every person has his or her motivation in different ways. People have certain ways of interacting, certain ways of working and certain ways of finishing the tasks. Along with this motivation is discipline. You need to be disciplined well in all aspects so that you can have a good motivation in dealing with the different personalities of people.

People are created differently with different purposes, personalities and motivations. This is why you are encouraged to understand the personality types so that when you deal with other people, you will know what to do. If you want to succeed in what you do, then you need to make sure that you understand the different types of people that surround you. This is the only way that you can get along with other people. If you understand them and their needs of belonging, you will surely be able to avoid conflicts. A lot of people are failing these days because they don't know how to understand and be understood by other people.

Benefits of Learning Personality Types

If you are a worker, it is so important that you understand all types of personality that your co-workers have. Knowing their personality will help you have a good workplace to work in. These are the benefits that you can get from understanding the personal types of people.

- You become a good team member and a team leader. You can recognize their strengths, preferences, weaknesses, blind spots and habit of mind and this will give you the idea that you need to make the most out of it and each of the person in your team. You will also know the type of work settings that you have and how you will handle everyone in your team. Being in a team and the leader at the same time takes so much responsibility because you are responsible for all your actions and your success as a team. You will become more productive because you are comfortable working with each other.

- This could also improve your communication with all your teams and everyone inside the workplace. You will understand their concerns, points of view and comments when it comes to your work. Another important thing is that when you notice something wrong with your workers especially in your workplace, you can easily figure out a way to fix it.

- When you understand the people and their personality type, you can also resolve the conflicts in your workplace faster and easily. You can be an effective peace maker and you can guide your workers accordingly. All the problems in your office will be easier to handle because you already know how to deal with your people.

- You can make a better decision as their leader. Being the leader of your team, it is your responsibility to make sure that everyone is doing their part. All your assignments and tasks will be done on time because you know how to manage your time with your decisions. As a leader, you are responsible for the success of your team. If you can guide them accordingly, then surely, you will reach the goals that you have set for your team and the company. Just make sure that you know their personality so that you will know how to approach your workers.

- You can also conduct better meetings so that all the workers in your team are always aware of the latest news in your work. They are always updated in everything that the company is doing. The workers have the right to know everything because they are also part of the company's success.

- You can manage the changes so that all your team members will be moving forward along with you. You and your team should move towards success. If you feel the success is going your way, it is just right that you also acknowledge the presence and hard work of your workers. This way, they will also be part of your success and they will not feel rejected and abandoned.

- Negotiate an ethical difference among the employees and become a good interviewer. Another important thing that you can also get from this is that you match the job requirements well. This way, you can assign the person who can do a certain job better.

These are the most important benefits that you can get from learning the different personality types of people. There are still other benefits that you can get from learning the different personality types. This is why it is important that you learn to understand people. By knowing the personality types, you can become a better person and succeed in your chosen career.

www.ingramcontent.com/pod-product-compliance
Lightning Source LLC
Chambersburg PA
CBHW031507210526
45463CB00003B/1117